EXPLORE!

FORCE

FLOODS

BY MONIKA DAVIES

Please visit our website, www.enslow.com. For a free color catalog of all our high-quality books, call toll free 1-800-398-2504 or fax 1-877-980-4454.

Cataloging-in-Publication Data

Names: Davies, Monika.
Title: Floods / Monika Davies.
Description: New York : Enslow Publishing, 2021. | Series: Force of nature | Includes glossary and index.
Identifiers: ISBN 9781978518421 (pbk.) | ISBN 9781978518438 (library bound)
Subjects: LCSH: Floods--Juvenile literature.
Classification: LCC GB1399.D383 2021 | DDC 551.48'9--dc23

Published in 2021 by
Enslow Publishing
101 West 23rd Street, Suite #240
New York, NY 10011

Copyright © 2021 Enslow Publishing

Designer: Katelyn E. Reynolds
Editor: Monika Davies

Photo credits: Cover, p. 1 2M media/Shutterstock.com; cover, pp. 1–48 (series art) Merfin/Shutterstock.com; pp. 5, 39 Scott Olson/Getty Images; p. 6 Tim Graham/The Image Bank/Getty Images Plus; p. 7 JAMES NIELSEN/AFP via Getty Images; p. 8 Chris Anson/Shutterstock.com; pp. 9, 41 courtesy of NASA; p. 10 peepy/Shutterstock.com; p. 11 Uladzimir Navumenka/Shutterstock.com; p. 12 serkan senturk/Shutterstock.com; p. 13 sirtravelalot/Shutterstock.com; p. 14 Ellen Bronstayn/Shutterstock.com; p. 15 Artisticco/Shutterstock.com; p. 17 JOSH EDELSON/AFP via Getty Images; p. 18 Jan Tove Johansson/The Image Bank/Getty Images Plus; p. 19 Theo Allofs/Corbis Documentary/Getty Images Plus; p. 20 Jerry Grayson/Heliflms Australia PTY Ltd/Getty Images; p. 21 Kyle Niemi/US Coast Guard via Getty Images; p. 22 Peter Cade/The Image Bank/Getty Images Plus; p. 23 DavorLovincic/E+/Getty Images; p. 24 MATT CAMPBELL/AFP via Getty Images; p. 25 VT750/Shutterstock.com; p. 27 Bill Brennan/ Perspectives/Getty Images Plus; p. 28 PHILIPPE LOPEZ/AFP via Getty Images; p. 29 Marcin Balcerzak/Shutterstock.com; p. 30 Lloyd Cluff/Corbis Documentary/Getty Images Plus; p. 31 Peter Hermes Furian/Shutterstock.com; p. 32 Baloncici/Shutterstock.com; p. 33 nootprapa/Shutterstock.com; p. 34 bubblea/Shutterstock.com; p. 35 Agence France Presse/Getty Images; p. 37 courtesy of the Library of Congress; p. 38 humphery/Shutterstock.com; p. 40 Asianet-Pakistan/Shutterstock.com; p. 42 Jenny Jones/Lonely Planet Images/Getty Images Plus; p. 43 Brien Aho/U.S. Navy via Getty Images; p. 44 Cameron Davidson/Corbis Documentary/Getty Images Plus.

Portions of this work were originally authored by Michael Portman and published as *Deadly Floods*. All new material in this edition is authored by Monika Davies.

All rights reserved. No part of this book may be reproduced in any form without permission in writing from the publisher, except by a reviewer.

Printed in the United States of America

Some of the images in this book illustrate individuals who are models. The depictions do not imply actual situations or events.

CPSIA compliance information: Batch #BS20ENS: For further information contact Enslow Publishing, New York, New York, at 1-800-542-2595.

Find us on

CONTENTS

Frightening Floods 4

The Water Cycle 6

Flood Facts 16

Flood Effects 26

Floods Throughout History 36

Glossary 46

For More Information 47

Index 48

WORDS IN THE GLOSSARY APPEAR IN **BOLD** TYPE THE FIRST TIME THEY ARE USED IN THE TEXT.

FRIGHTENING FLOODS

Water is a necessary ingredient for human survival. In fact, adult human bodies are made of up to 60 percent water! Water helps the human body flush out waste, move oxygen around, encourage body cells to grow, and much more. People also use water to grow their crops, flush their toilets, and cook meals. Water plays an important role in people's lives, but it can also quickly turn into a deadly force of nature.

A flood is a large stream of rushing water that covers land that's normally dry. Floods are the most common weather-related natural **disaster** to occur worldwide. This makes them a force of nature that affects millions of people.

IN SPRING 2011, HEAVY RAINS AND MELTING SNOW CAUSED THE MISSISSIPPI RIVER, THE LARGEST RIVER IN THE UNITED STATES, TO EXPERIENCE MASSIVE FLOODING. THE FLOODWATERS FORCED THOUSANDS OF PEOPLE IN SEVERAL STATES TO FLEE THEIR HOMES.

THE WATER CYCLE

Flooding can happen almost anywhere in the world. Even areas that aren't close to rivers, streams, lakes, or oceans can experience flooding. When rivers or lakes overflow, dams break, or ocean waves crash far onto shore, the results can be **devastating**.

IN THE UNITED STATES, FLOODS CAUSE ABOUT $8 BILLION WORTH OF **DAMAGE** AND OFTEN KILL OVER 100 PEOPLE EACH YEAR. WORLDWIDE, FLOODS ARE RESPONSIBLE FOR AROUND $40 BILLION WORTH OF DAMAGE.

Water is one of the most important elements on Earth. Every living thing depends on water for life. People use water for drinking, cooking, cleaning, farming, and having fun. Ships and boats travel on the world's oceans, rivers, and lakes. With all its important uses, it can be easy to forget just how powerful and destructive water can be.

Since water is a necessary part of life, some of the most valuable and important land is near water. Many cities and towns have been built near sources of water. Millions of people live in homes that are very close to rivers and coastlines. Unfortunately, this has resulted in millions of deaths and trillions of dollars in damage from floods.

DESTRUCTIVE:
CAUSING SOMETHING TO BE DESTROYED OR RUINED

PORTLAND, OREGON, IS ONE EXAMPLE OF A CITY LOCATED NEAR WATER. THE WILLAMETTE RIVER FLOWS BETWEEN THE CITY'S EAST AND WEST COMMUNITIES AND ALSO SERVES AS A CHANNEL FOR LARGE SHIPS.

WATERY PLANET

Around 71 percent of Earth is covered in water, and most of that water is contained in Earth's oceans! Approximately 96.5 percent of Earth's water can be found in our oceans. The amount of water on Earth has remained the same for millions of years. The only thing that's changed is where the water is located. For example, some bodies of water have dried up, and some have formed where none existed before. Glaciers and ice sheets have melted and sent water flowing into rivers and oceans.

VIEWED FROM SPACE, EARTH LOOKS BLUE! THAT'S BECAUSE OF HOW MUCH OF THE PLANET IS COVERED IN WATER.

Water is continuously moving on, above, and below Earth's surface. This movement is called the water cycle. The water cycle starts when heat from the sun turns water into a gas called water vapor. This process is called evaporation. The water vapor then rises into the air, where cooler temperatures turn it back into a liquid (or even ice). This process is called condensation. The liquid then combines with dust and other particles. These particles form clouds.

CONDENSATION IS EASY TO SEE ON WINDOWS, AS SHOWN HERE, OR ON THE OUTSIDE OF A GLASS OF COLD WATER ON A WARM DAY.

EXPLORE MORE

MOST OF THE MOISTURE IN OUR ATMOSPHERE COMES FROM WATER EVAPORATING FROM OCEANS, LAKES, RIVERS, AND SEAS. IN FACT, ALMOST 90 PERCENT OF THE MOISTURE IN THE ATMOSPHERE COMES DIRECTLY FROM EVAPORATION.

WATER EVAPORATION FROM THE SURFACE OF OCEANS IS A KEY DRIVER OF THE WATER CYCLE. AROUND 86 PERCENT OF WATER EVAPORATION INTO THE ATMOSPHERE COMES FROM EARTH'S OCEANS.

Particles in the clouds grow larger and heavier. Then the water falls back to Earth as rain, sleet, hail, or snow. This is called precipitation. Finally, the water that falls back to Earth evaporates, and the cycle begins again.

PARTICLE:
A VERY SMALL PIECE OF SOMETHING

SEEING PRECIPITATION OCCUR IS WATCHING PART OF THE WATER CYCLE!

WHAT HAPPENS WHEN WATER FALLS?

Several different things can happen to water that falls as precipitation. Some of the water lands in oceans, rivers, and lakes. Some of it ends up on land. Sometimes, the ground absorbs the water. Plants then take up some of the water in the ground. The rest may travel through the ground until it empties into a body of water. Water that doesn't get absorbed travels across land until it reaches streams or rivers. Usually, nature does a good job of keeping the water cycle balanced. However, if there's too much water at certain points in the cycle, flooding can occur.

FLOODING DOESN'T ALWAYS OCCUR ON A LARGE SCALE. IT CAN HAPPEN RIGHT IN YOUR BACKYARD!

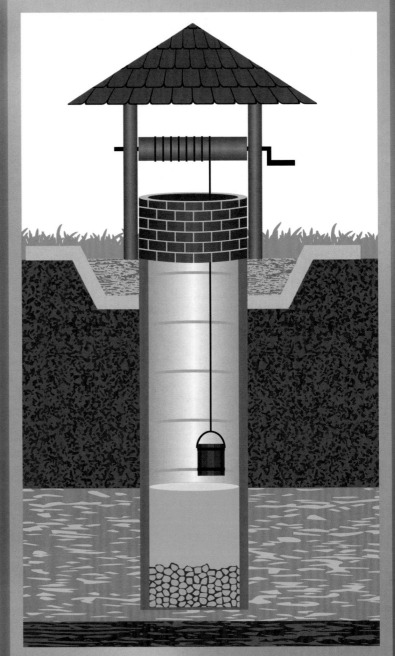

WATER THAT STAYS DEEP UNDERGROUND IS OFTEN KNOWN AS GROUNDWATER. PEOPLE DRILL WELLS IN ORDER TO BRING GROUNDWATER BACK TO THE SURFACE FOR DRINKING AND OTHER USES.

EXPLORE MORE

NOT ALL THE WATER THAT GETS ABSORBED BY THE GROUND ENDS UP IN RIVERS. SOME WATER FINDS ITS WAY DEEP UNDERGROUND INTO CRACKS AND SPACES IN SOIL, SAND, AND ROCK.

ABSORB:
TO TAKE IN

14

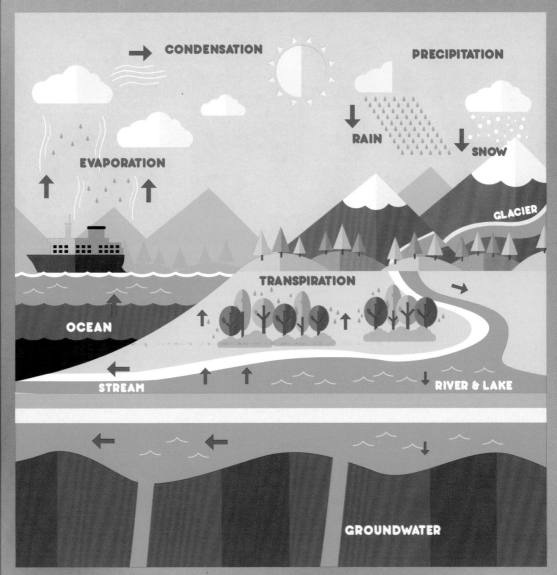

FLOOD FACTS

Floods occur for many reasons. Sometimes, a river overflows its banks. Heavy rainfall can dump more water into a river than it can hold. Melting snow, especially when combined with rain, can cause water levels to rise rapidly.

Other times, something blocks the steady flow of river water. Man-made dams and natural dams—such as blocks of ice, beaver dams, and logjams—can stop or greatly reduce the flow of water. When this happens, the land behind the dam will flood if water levels rise too high. If the dam breaks, a great rush of water will speed down the river all at once.

IN FEBRUARY 2017, A HOLE WAS DISCOVERED IN THE MAIN SPILLWAY, OR CHANNEL FOR EXTRA WATER, OF CALIFORNIA'S OROVILLE DAM. THE DAM WAS CLOSED BUT RAINWATER BEGAN TO FLOW THROUGH THE SPILLWAY. OFFICIALS WORRIED THIS WOULD CAUSE A FLOOD AND CALLED FOR AN EVACUATION OF RESIDENTS LIVING DOWNSTREAM.

OVERFLOW:
TO FLOW OVER THE TOP OF SOMETHING

 GLACIERS WORLDWIDE ARE MELTING. THIS HAS LED TO AN INCREASE IN SEA LEVELS, WHICH IN TURN INCREASES THE RISK OF FLOODING IN CITIES THAT ARE LOCATED NEAR OCEANS.

THIS PHOTOGRAPH SHOWS THE SIMPSON DESERT IN QUEENSLAND, AUSTRALIA, FLOODED WITH WATER. AS STRANGE AS IT MAY SOUND, MORE PEOPLE DROWN IN THE DESERT THAN DIE OF THIRST!

EXPLORE MORE

IT MAY BE HARD TO BELIEVE, BUT DESERTS HAVE SOME OF THE BIGGEST FLOODS! DESERT SAND ISN'T GOOD AT ABSORBING WATER QUICKLY. THUNDERSTORMS CAN FILL DRY RIVERBEDS AND LAKE BEDS IN A MATTER OF MINUTES.

Floods may also occur when the ground is saturated. Saturation means the ground is too wet to absorb any more water. To understand this, imagine pouring water onto a sponge. The sponge will absorb the water at first. However, once the sponge becomes completely wet, or saturated, water runs off it.

Some types of ground do a better job of absorbing water than others. Soil mixed with sand can absorb lots of water, while soil rich in clay absorbs much less. Just like a sponge, all soils reach a point where they become saturated. Extra water can result in a flood.

HURRICANE KATRINA HIT NEW ORLEANS, LOUISIANA, IN AUGUST 2005. LEVEES, OR **BARRIERS** BETWEEN THE OCEAN AND THE CITY, GAVE WAY. THIS CAUSED CONSIDERABLE FLOODING IN THE LOW-LYING AREAS. AROUND 80 PERCENT OF THE CITY WAS FLOODED.

MAN-MADE DRAINAGE

While constructing cities, towns, and roads, people have covered large areas with buildings, stones, and concrete. Since these types of materials, or matter from which something is made, can't absorb much water, cities and towns have drainage systems. A drainage system is a system of pipes or channels that carry water away from a place. However, city drainage systems sometimes fail. The devastating 2005 flood of New Orleans resulting from Hurricane Katrina is one example. The drainage system in New Orleans wasn't built to carry such an immense amount of water.

NEW ORLEANS, LOUISIANA
AUGUST 29, 2005

Soil that's frozen doesn't absorb water very well. If snow melts or it rains before frozen ground has thawed, water that would usually be absorbed can contribute to a flood. Human activity can also affect how the ground absorbs water. Soil that's been plowed, such as farmland, usually isn't very good at soaking up water.

Not all floods are alike. Some floods develop slowly and steadily. Others develop quickly, sometimes without warning. Slow, or gradual, floods may take several days to become dangerous. Gradual flooding usually means that people have enough time to get to safety. However, these floods can still cause plenty of damage.

COASTAL FLOODING IS ANOTHER TYPE OF FLOOD THAT OCCURS WHEN SEAWATER FLOODS A COASTLINE AND ITS SURROUNDING AREAS. THIS OFTEN HAPPENS WHEN THERE'S A STORM SURGE, OR A RISING OF THE SEA DUE TO A STORM.

THAW:
TO NO LONGER BE FROZEN OR TO CAUSE SOMETHING TO NO LONGER BE FROZEN

A SPRING THAW CAN LEAD TO SPRING FLOODS. AS TEMPERATURES GET WARMER, SNOW CAN BEGIN TO MELT AT A FAST PACE. THE RESULTING WATER FROM MELTED SNOW CAN LEAD TO FLOODING.

RAGING WATER

Floods that happen quickly are called flash floods. Flash floods are the most dangerous type of flood. Flash floods can turn a calm stream or dry road into a raging river in a matter of minutes. Since flash floods happen with little or no warning, people usually don't have time to prepare for them. Flash floods can bring waves of water that are over 30 feet (9 m) high. In the United States, most flood-related deaths are caused by flash floods. Over half of flash flood-related deaths involve people in cars and trucks. Even low water can push cars off roads, causing accidents that kill the passengers in the vehicle.

IN EARLY MARCH 1997, HEAVY RAINS POURED DOWN ON KENTUCKY AND INDIANA, RESULTING IN MULTIPLE FLASH FLOODS THAT OVERTOOK AND DESTROYED BUILDINGS IN THE AREA. OVER 14,000 HOMES WERE DESTROYED IN THE FLOODS.

EXPLORE MORE

FLASH FLOODS USUALLY OCCUR DUE TO SEVERAL ELEMENTS. HOWEVER, THE TWO MOST IMPORTANT ELEMENTS ARE THE INTENSITY OF RAINFALL AND HOW LONG THE RAIN COMES DOWN. A FLASH FLOOD MAY OCCUR IF RAIN IS COMING DOWN WITH GREAT FORCE IN A SHORT PERIOD OF TIME.

INTENSITY:
THE AMOUNT OF FORCE SOMETHING HAS

FLOOD EFFECTS

Floods are a powerful force of nature that can have devastating consequences. Cars, trees, bridges, and people can be swept away by the rushing currents of water. Water can rip buildings from their bases, causing them to fall down. The resulting **debris** may be carried over large areas.

Floods that cover farmland can destroy many fields of crops and drown livestock and other animals. Widespread flooding may cause food shortages too.

TSUNAMI:
A HUGE WAVE OF WATER CREATED BY AN UNDERWATER EARTHQUAKE OR VOLCANO

IN DECEMBER 2004, A TSUNAMI IN THE INDIAN OCEAN CAUSED TERRIBLE FLOODING IN MANY COUNTRIES, INCLUDING INDONESIA AND THAILAND. MORE THAN 225,000 PEOPLE DIED.

In addition to destroying property, floods can also spread sickness and disease. Floodwaters are often filled with mud, sewage, and poisonous chemicals. Even if a house is still standing after a flood, dirty water can cover everything, making it an unhealthy place to live.

SEWAGE:
WASTE MATTER THAT IS TRANSFERRED OUT OF HOMES AND OTHER BUILDINGS THROUGH A SERIES OF PIPES

IT TAKES ONLY 6 INCHES (15 CM) OF RUSHING WATER TO KNOCK A PERSON OFF THEIR FEET. THIS IS WHY IT'S IMPORTANT DURING A FLOOD TO STAY WHERE YOU ARE AND NOT TRY TO CROSS MOVING FLOODWATERS.

EXPLORE MORE

FLOODWATER DOESN'T HAVE TO BE VERY DEEP TO BE DESTRUCTIVE. IT MAY NOT SEEM LIKE MUCH, BUT JUST 2 FEET (61 CM) OF WATER CAN HAVE ENOUGH FORCE TO CARRY A CAR.

COMING INTO CONTACT WITH FLOODWATERS CAN SOMETIMES INCREASE YOUR CHANCE OF INFECTION, OR A SICKNESS CAUSED BY GERMS. WATERBORNE ILLNESSES CAN **CONTAMINATE** DRINKING WATER FOLLOWING FLOODING.

However, not every flood is a disaster. Floodwaters can leave behind important **nutrients** that are good for soil. One example is the flooding of the Nile River in Egypt. For thousands of years, Egyptians depended on the seasonal flooding of the Nile to **fertilize** their soil. Farmers looked forward to the heavy rains that caused the Nile to flood.

In 1970, a dam was built on the Nile River to control the water flow. Water is let out as it's needed. This has allowed the planting season to last all year, resulting in more food.

THE ASWAN HIGH DAM CROSSES THE NILE RIVER AND ALLOWS CONTROL OVER THE ANNUAL NILE FLOOD. FLOODWATERS GATHER IN THE DAM AND CAN BE RELEASED, OR FREED, WHEN NEARBY CROPS NEED WATERING.

NILE NUTRIENTS

The Nile River is one of the world's longest rivers, spanning 4,100 miles (6,600 km). It flows from south to north and empties into the Mediterranean Sea. The Nile River delta is situated between Cairo and the Mediterranean Sea. A delta is land shaped like a triangle at the mouth of a river. As the Nile River flowed into the sea, it left behind nutrient-rich silt in the delta. The majority of Egypt's food is grown in the Nile River delta, including beans and wheat. Unfortunately, much of the silt that used to flow toward the delta is now instead caught behind the Aswan High Dam. This has caused the delta to begin decreasing in size.

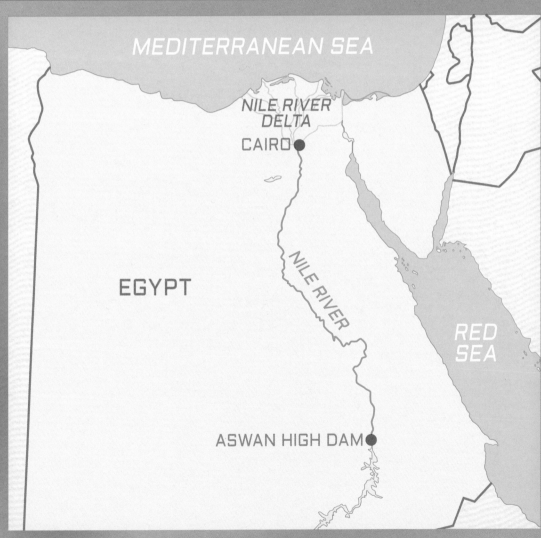

THE NILE WAS AN IDEAL PLACE TO SETTLE IN ANCIENT TIMES. IT STILL IS TODAY! ABOUT 95 PERCENT OF EGYPTIANS LIVE JUST A FEW MILES FROM THE NILE.

For thousands of years, people have been trying to control floods. They build dams to control the flow of water. They build levees to keep rivers from overflowing. They build tunnels and concrete channels to move water from one place to another. Sometimes, the efforts are successful. Other times, they have disastrous effects.

If dams and levees aren't tall enough, rising floodwaters can overflow them. If they're old or poorly made, water pressure can cause them to break. Dams, levees, and channels may prevent flooding in one area but direct water into other areas, causing flooding there.

MAN-MADE LEVEES ARE OFTEN MADE OF SOIL, SAND, OR ROCKS PILED UP ALONG A BODY OF WATER. LEVEES MAY ALSO BE BUILT USING WOODEN OR PLASTIC BLOCKS.

CHANNEL:
A LONG, NARROW AREA WHERE WATER FLOWS

THE HOOVER DAM IS ONE OF THE MOST FAMOUS DAMS IN THE UNITED STATES. IT STRETCHES ACROSS THE COLORADO RIVER BETWEEN ARIZONA AND NEVADA. THE DAM IS RESPONSIBLE FOR CONTROLLING FLOODING ALONG THE RIVER.

EXPLORE MORE

THE HOOVER DAM IS THE LARGEST DAM IN THE UNITED STATES. IT ALSO CREATES HYDROELECTRICITY. THERE ARE 17 TURBINES CREATING HYDROELECTRIC POWER FOR PEOPLE TO USE IN NEVADA, ARIZONA, AND CALIFORNIA.

WATER POWER

Electricity created with water is called hydroelectricity. Rushing water spins large metal blades called turbines that are connected to machines called generators. The generators create electricity that's sent to homes and businesses. Hydroelectricity is less harmful to Earth than fuels such as gas and coal. The Three Gorges Dam in China is the world's largest hydroelectric dam. It was built both to produce electricity and to reduce flooding on the Yangtze River. Construction began in 1994, and the main dam was completed in 2006. In 2011, Chinese officials acknowledged the dam had **displaced** many people and also caused environmental damage.

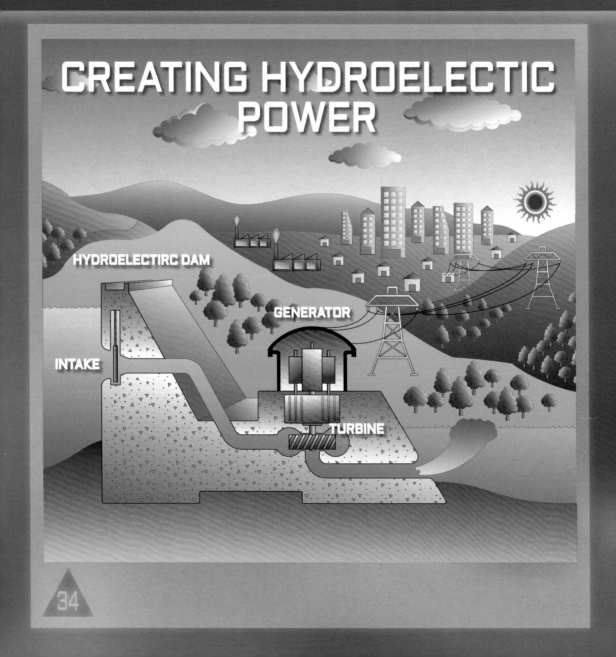

THE THREE GORGES DAM IS LOCATED ON THE YANGTZE RIVER. THE WORLD'S LARGEST HYDROELECTRIC DAM IS TRULY GIGANTIC IN SIZE, MEASURING 594 FEET (181 M) HIGH AND 7,770 FEET (2,335 M) LONG.

FLOODS THROUGHOUT HISTORY

Floods have caused mass devastation throughout history and across many countries. One of the worst floods in U.S. history occurred on May 31, 1889, when a dam broke near Johnstown, Pennsylvania.

After heavy rains, townspeople were warned of the coming flood. Sadly, most people thought the flood would be like others they had experienced. However, witnesses said their dam "just moved away," allowing water to destroy much of the town.

IN THE JOHNSTOWN FLOOD, 2,209 PEOPLE DIED AND 1,600 HOMES WERE DEMOLISHED. AT THE TIME, THE FLOOD CAUSED $17 MILLION IN PROPERTY DAMAGE.

China has had some of the worst floods in history. Many of those floods took place on Asia's longest river, the Yangtze. In 1931, heavy rains caused the river to flood a massive area, including the nearby cities of Nanjing and Wuhan. That particular flood was responsible for 300,000 deaths and left more than 40 million people without a home.

EXPLORE MORE

ACCORDING TO A 2016 UNITED NATIONS (UN) REPORT, FLOODS AFFECTED THE LIVES OF AROUND 2.3 BILLION PEOPLE FROM 1995 TO 2015. IN THAT SAME TIME FRAME, AROUND 157,000 PEOPLE DIED BECAUSE OF FLOODS.

IN 2017, TORRENTIAL RAINS POURED DOWN ON SOUTHERN CHINA, CAUSING WATER LEVELS TO RISE IN THE YANGTZE RIVER. THE HEAVY RAINFALL LED TO SEVERE FLOODING ACROSS MULTIPLE PROVINCES IN CHINA.

2019 MIDWESTERN FLOODS

In 2019, the midwestern United States was swamped with massive floods. This extensive flooding continued over many months, causing deaths, destruction in several states, and delays in planting crops. Torrential downpours in the Midwest began in March 2019, hitting ground that was still frozen. Frozen ground can't absorb water well, which led to higher and higher water levels. **Experts** have also pointed to climate change as a factor in the increased rainfall. Four rivers—the Mississippi, the Illinois, the Missouri, and the Arkansas—all experienced flooding. Nearly 14 million people were affected by the record flooding.

THE MISSISSIPPI RIVER'S WATER LEVEL WAS SO HIGH IN MAY 2019 THAT IT FLOWED OVER THIS ROADWAY AT THE BORDER OF MISSOURI AND ILLINOIS.

TORRENTIAL: COMING IN A LARGE, QUICK FLOW OR STREAM

In summer 2010, Pakistan suffered one of the worst natural disasters in recent history. Monsoon rains poured down on large areas of the country, and flooding ensued. Houses and crops were destroyed, leaving millions of people without homes and access to an **adequate** food supply. Nearly 20 million people were affected in the flood.

FOR MANY YEARS PRIOR TO THE 2010 FLOODING IN PAKISTAN, A LARGE NUMBER OF TREES, SHRUBS, AND PLANTS WERE CUT DOWN OR REMOVED THROUGHOUT THE COUNTRY. THIS MADE THE GROUND LESS ABLE TO SOAK UP MOISTURE, WHICH INCREASED THE RISK OF FLOODS.

THE EFFECTS OF CLIMATE CHANGE

Climate change is the long-term change in Earth's climate, caused partly by human activities such as burning oil and natural gas. This has directly led to an increase in global temperatures. In particular, climate change has led to temperatures rising in Earth's atmosphere. The warmer the atmosphere, the more moisture it can hold. This directly leads to heavier precipitation in comparison to the past. Heavier precipitation can increase the **potential** for more devastating flooding. Climate change has also led to a greater number of hurricanes. The strongest hurricanes bring heavier rainfall, which can also lead to destructive flooding.

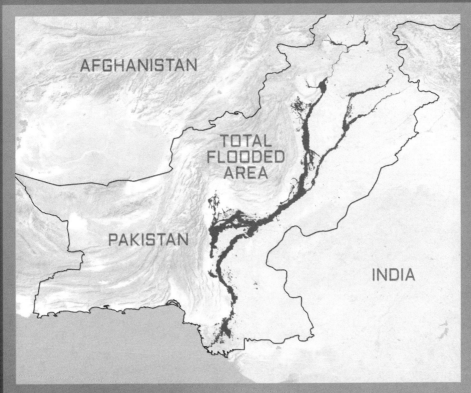

THE FLOODING IN PAKISTAN COVERED AT LEAST 14,390 SQUARE MILES (37,280 SQ KM) BETWEEN JULY AND SEPTEMBER 2010, THOUGH NOT ALL THAT AREA HAD FLOODING AT THE SAME TIME.

MONSOON:
A SEASONAL CHANGE IN WIND DIRECTION RESULTING IN A CHANGE IN PRECIPITATION

Technology has developed so floods can often be **predicted** and prepared for. The National Weather Service (NWS) monitors rainfall across the United States. The NWS uses this data to predict potential floods and issue warnings. Local police, fire departments, and rescue forces are also prepared for these natural disasters and are usually the first on the scene when there's a flood. If necessary, U.S. armed forces may be called in to help as well. It's a good idea to carefully follow the instructions of rescue workers as well as local health departments during and after a flood.

TECHNOLOGY:
TOOLS, MACHINES, OR WAYS TO DO THINGS THAT USE THE LATEST DISCOVERIES TO FIX PROBLEMS OR MEET NEEDS

FLOODS ARE ALSO PREDICTED BY MONITORING THE CHANGING LEVEL OF WATER IN A RIVER. THIS HELPS EXPERTS PREDICT WHETHER THE RIVER WILL OVERFLOW, AS WELL AS HOW SEVERE THE FLOODING COULD BE.

FOLLOWING HURRICANE KATRINA, THE U.S. NATIONAL GUARD, ARMY, NAVY, MARINE CORPS, AND COAST GUARD WERE CALLED IN TO HELP.

EXPLORE MORE

FLOODING HAPPENS IN EVERY U.S. STATE, AS FLOODING OCCURS WHEREVER THERE'S RAINFALL. IN FACT, FLOODS ARE A DEADLIER THREAT THAN TORNADOES OR HURRICANES! MORE PEOPLE DIE FROM FLOODS IN THE UNITED STATES THAN ANY OTHER NATURAL DISASTER.

A flood can devastate a city or country in a very short amount of time. Floods can happen almost anywhere and are the most common weather-related disaster that affects people worldwide. However, although floods present many dangers and risks, there are a number of ways people can prepare and ready themselves to stay safe when a flood strikes.

IN THE SUMMER OF 1993, FLOODING OCCURRED IN NINE U.S. STATES AS A RESULT OF RECORD RAINFALL FROM JUNE TO AUGUST. THE MISSISSIPPI RIVER, SHOWN HERE, AND THE MISSOURI RIVER BOTH CAUSED FLOODING.

FLOOD SAFETY

- DON'T TRY TO CROSS MOVING WATER ON FOOT OR IN A CAR.

- AVOID CROSSING BRIDGES THAT RUN OVER RAPIDLY MOVING WATER.

- ASK YOUR FAMILY TO SIGN UP FOR YOUR LOCAL ALERT SYSTEM, THE EMERGENCY ALERT SYSTEM (EAS), OR THE NATIONAL OCEANIC AND ATMOSPHERIC ADMINISTRATION (NOAA) WEATHER RADIO.

- LISTEN TO TV AND RADIO REPORTS TO STAY INFORMED OF A FLOOD'S PROGRESS. (BATTERY-POWERED RADIOS ARE BEST IN CASE ELECTRICITY IS CUT OFF.)

- IF THERE'S DANGER OF A FLASH FLOOD IN YOUR AREA, BE SURE TO GET TO HIGHER GROUND.

- EVEN IF THE WATER ISN'T FLOWING VERY FAST, IT MAY CONTAIN THINGS THAT WILL MAKE YOU SICK. DON'T WADE THROUGH FLOODWATERS.

- AFTER A FLOOD, RETURN TO YOUR HOUSE ONLY AFTER YOU'VE BEEN TOLD IT'S SAFE.

GLOSSARY

adequate Enough to meet a need.

barrier Something that makes progress hard.

contaminate To pollute something.

damage Harm; also, to cause harm.

debris Pieces of something that has been destroyed or broken.

devastating Causing widespread damage.

disaster An event that causes much suffering or loss.

displace To require people to leave the place where they live.

expert Someone who knows a great deal about something.

fertilize To add something to soil to increase its ability to grow plants.

hurricane A powerful storm that forms over water and causes heavy rainfall and high winds.

nutrient Something a living thing needs to grow and stay alive.

potential Possibly existing.

predict To guess what will happen in the future based on facts or knowledge.

FOR MORE INFORMATION

BOOKS

DK Publishing. *Eyewitness Weather*. New York, NY: DK Publishing, 2016.

Elkins, Elizabeth. *Investigating Floods*. North Mankato, MN: Capstone Press, 2017.

Gunderson, Jessica. *Carrie and the Great Storm: A Galveston Hurricane Survival Story*. North Mankato, MN: Stone Arch Books, 2019.

Spradlin, Michael P. *Nile Chaos: A 4D Book*. North Mankato, MN: Stone Arch Books, 2018.

WEBSITES

Floods
www.ready.gov/floods
Learn more about how to stay safe during and after a flood.

How Floods Work
science.howstuffworks.com/nature/natural-disasters/flood.htm
Find out more about the causes and dangers of floods.

Rain & Floods
www.weatherwizkids.com/weather-rain.htm
Discover what causes rain and leads to a flood.

Publisher's note to educators and parents: Our editors have carefully reviewed these websites to ensure that they are suitable for students. Many websites change frequently, however, and we cannot guarantee that a site's future contents will continue to meet our high standards of quality and educational value. Be advised that students should be closely supervised whenever they access the internet.

INDEX

A
Aswan High Dam, 31

C
causes of floods, 6, 16, 20, 25, 32, 41
China, 34, 38
climate change, 39, 41
cloud formation, 10
condensation, 10

D
damage from floods, 8, 22, 26, 36, 38, 40, 44
dams, 6, 16, 30, 31, 32, 33, 34, 36
deaths from floods, 8, 24, 38, 39, 43
deserts, 19
drainage systems, 21

E
evaporation, 10, 11, 12

F
flash floods, 24, 25
floodwaters, 28, 29, 30, 32

G
gradual flooding, 22

H
Hoover Dam, 33
Hurricane Katrina, 21
hydroelectricity, 33, 34

J
Johnstown, Pennsylvania, flood, 36

L
levees, 32

M
Midwest, U.S., floods of 2019, 39

N
Nile River, 30, 31

P
Pakistan, 40
precipitation, 12, 13, 41
predicting floods, 42

S
saturation, 20

T
Three Gorges Dam, 34

W
water absorption, 13, 14, 19, 20, 21, 22, 39
water cycle, 10–12, 13